GU00640715

Chelation Therapy
for Cardiovascular Health

Second Edition
Revised and Updated

C. M. Hawken

WOODLAND PUBLISHING
TM

Copyright © 2007 C. M. Hawken

All rights reserved. No part of this publication may be reproduced, stored in a retrieval system, or transmitted in any form without the prior permission of the copyright owner.

For ordering information and bulk order discounts, contact:
Woodland Publishing, 448 East 800 North, Orem, UT 84097
Toll-free telephone: (800) 777-BOOK

Please visit our Web site: www.woodlandpublishing.com

Note: The information in this book is for educational purposes only and is not recommended as a means of diagnosing or treating an illness. All matters concerning physical and mental health should be supervised by a health practitioner knowledgeable in treating that particular condition. Neither the publisher nor the author directly or indirectly dispenses medical advice, nor do they prescribe any remedies or assume any responsibility for those who choose to treat themselves.

A cataloging-in-publication record for this book is available from the Library of Congress.

ISBN: 978-1-58054-431-3

Printed in the United States of America

Contents

Cardiovascular Disease and Chelation Therapy 5

What Is Chelation Therapy? 6

Oral Chelation Therapy 7

Arteriosclerosis: Cause and Mechanism 9

Standard Treatments for Heart Disease 14

History of EDTA and Chelation Therapy 15

Research on Chelation Therapy 17

Interview with Dr. Garry Gordon 19

Other Benefits of EDTA Chelation Therapy 20

What Are the Costs of Chelation Therapy? 21

How Safe Is Chelation Therapy? 21

References 23

Cardiovascular Disease and Chelation Therapy

More people in the United States die each year from heart-related diseases than from any other cause. These diseases, collectively called cardiovascular disease, include arteriosclerosis (hardening of the arteries); heart attack (myocardial infarction); arrhythmias (irregular heart beat); stroke; hypertension (high blood pressure); congestive heart failure, and rheumatic heart disease. Approximately one million people die each year from these diseases, nearly half of all the people who die in the United States. This is more than double the number who die from cancer, and nearly twenty-five times the number of those who die from AIDS.

How many people suffer from some form of heart disease? The statistics are frightening: nearly one in every four Americans has at least one form of cardiovascular disease. Having heart disease certainly leads to shortened life expectancy, varying degrees of physical suffering, missed work time and overall disability. In fact, a large portion of all disabilities is comprised of some type of cardiovascular disease.

In response to the growing number of people with heart disease, there is sound and ample evidence that shows that chelation therapy—an extremely safe, effective and nontoxic procedure—can be highly effective in preventing and reversing the debilitating effects of various cardiovascular conditions. This procedure can reduce arterial plaque and lesions, increase elasticity of blood vessels, improve overall circulation, and decrease free radical damage to the cardiovascular system. One does not need to be a genius to deduce how much time, suffering and money can be saved if, indeed, chelation therapy is capable of these things.

Heart disease occurs in such an epidemic proportion in our country that various and sometimes disturbing questions must be asked: Why is heart disease so common? What can be done to prevent it? And why are invasive, costly, and mostly ineffective medical procedures the only methods available to treat cardiovascular disease? The following sections will provide answers to these questions, and ultimately point to chelation therapy as an effective and safe remedy to many types of heart disease.

What Is Chelation Therapy?

Chelation therapy consists of intravenous injections of a number of agents, but mostly of a synthetic amino acid called ethylene diamine tetraacetic acid (EDTA). Upon being introduced intravenously into the bloodstream, EDTA binds or "chelates" with certain minerals and substances that are present in the bloodstream and in the walls of blood vessels, and disposes of them via the urine. The most common mineral targeted by EDTA chelation therapy is calcium, whose irregular metabolism is thought to have a major role in spasms and constriction of the arteries. Calcium is also thought to be a principal player in circulatory impairment and other disorders of the circulatory system. An overabundance of calcium can also cause blood vessels to become calcified and less elastic, not fully allowing the blood to reach and nourish the intended parts of the body. Calcium is also part of the composition of arterial plaque, one of the main culprits in heart attacks and many forms of heart disease. Removal of calcium in the bloodstream and in arterial plaque could prove invaluable in not only preventing all manners of circulatory and coronary disorders, but in reversing their effects as well.

How Does EDTA Chelation Work?

EDTA therapy has multiple mechanisms of action that positively affect arterial plaque formation and cell membrane function. As stated earlier, EDTA binds or chelates with the overabundant or offending mineral. Technically, chelation is defined as "the incorporation of a metal ion into a heterocyclic ring structure." It is this binding action that is so valuable in removing an intended substance from the circulatory system or other parts of the body. The substance is then disposed of, mainly via urine. Multiple studies have shown that EDTA is successful in removing calcium from blood, arterial plaque, arterial walls and other body areas. (As far as removing needed calcium from bones and related areas, there are studies that show EDTA does not bind with significant amounts.)

Other studies have shown ETDA's ability to effectively fight free radical scavengers, which can be very important in maintaining coronary and vascular health. There is data pointing to an increase in free radical activity after primary coronary angioplasty in acute myocardial infarction. Because chelating agents can offer control free radical activity, they appear to be useful agents to improve the outcome in patients with coronary artery disease (Patterson, 1528–35). Also, it appears that free radicals cause the linings of the blood vessel walls to become irritated, causing the body to send calcium to aid in the healing of the irritated area. As a result, the calcium binds with cholesterol, fats, scar tissue and other debris to build a debilitating coating over the affected area. It is this buildup that is referred to as arterial plaque.

There is another reason for EDTA being an ideal mediation for coronary and vascular disease. Recently EDTA has become one of a relatively new group of medicines known as calcium channel blockers. These medicines curb the flow of calcium and other minerals into the muscular coatings of the arteries, hence decreasing the occurrence of arterial spasms and ultimately relieve chest pain, irregular heartbeats, and other symptoms associated with coronary and circulatory disorders. EDTA is different, however, from the other calcium channel blockers. These drugs affect only the spasm aspect of calcium overload, while EDTA relieves both the spasms and the calcification of the arteries for a wider range of relief of coronary and circulatory disorders, particularly atherosclerosis (Walker 1982, 11).

Oral Chelation Therapy

Oral chelation therapy is similar to intravenous chelation therapy in its action, but is somewhat different in its mode of application and potency. The same principal that enables intravenous chelators to bind or chelate a metallic ion also works for oral chelation products. With IV chelation therapy, the chelating agent is injected directly into the bloodstream, the cardiovascular system being the principal target. With oral chelation, the chelating products are taken orally, digested, and assimilated so that their chelating effects apply to all parts of the body—its systems, organs, tissues and cells. Dr. Morton Walker defines the oral chelation process as "a prolonged process of chemical detoxification of the most inner recesses of one's physiology" (Walker 1997, 152).

Is Oral Chelation the Answer to Cardiovascular Disease?

Most experts agree that oral chelation products are not as speedy or effective in reversing the effects of arteriosclerosis, atherosclerosis and other cardiovascular disorders. Because of their mode of application (they must pass through the digestive process before reaching the bloodstream), they are not nearly as potent as the IV chelation products such as EDTA. Therefore, oral chelation is not the first choice of treatment for already advancing cardiovascular disease.

However, despite its supposed drawbacks of time effectiveness and potency, oral chelation therapy has been shown to reverse conditions of the cardiovascular system. Dr. Garry Gordon notes that he has seen persons with cardiovascular disease who, for various reasons did not or could not undergo the IV chelation therapy, began taking oral chelation products and eventually experienced dramatic improvement in their condition. He notes "I've had doctor friends who wouldn't take the IV at first, but who now are on oral EDTA, and are able to pass a treadmill stress test that they couldn't pass for five years. I've had lots of good things happen with oral EDTA–based supplement programs" (Gordon Interview, 15). Dr. Ward Dean also supports the use of oral chelation products specifically for cardiovascular and other "irreversible" conditions. He states, "I, too, have seen and heard of dramatic clinical improvements in very debilitated patients who had used only oral chelation as a treatment" (Dean, 13).

Oral chelation has other benefits as well. Because oral chelation products move far beyond just the body's cardiovascular system, they are able to effectively detoxify the body of various unwanted substances. For instance, Dr. Walker says that the most common benefit of oral chelation is the suppression of free radicals, which bring about all kinds of detrimental changes in the body, including that of irritating the lining of blood vessel walls (Walker 1997, 153).

Ingredients of Chelation Products

Many oral chelation products contain EDTA, the best-known chelator, as well as other recognizable chelating agents. Garlic is a popular ingredient known for its ability to inhibit blood platelet aggregation and clotting, lower serum cholesterol levels, and aid digestion. Most

B vitamins (including B1, B3, B6, B12, and pantothenic acid) are included in oral formulations. They perform a variety of functions, including neutralizing free radicals, promoting red blood cell formation and metabolism, and lowering serum cholesterol levels. Vitamin A promotes cell growth and repairs damaged cell membranes. Vitamin E prevents oxidation of fat compounds and is an anticoagulant. Magnesium (often taken with calcium) assists in calcium and potassium uptake and protects the arterial lining. Copper is needed for various cardiovascular functions, including the formation of hemoglobin, elastin and red blood cells. Pectin helps fight free radicals, removes unwanted toxins, and lowers cholesterol. Amino acids (L-cysteine, L-methionine and others) are powerful detoxifiers and antioxidants, and aid in the prevention of fat buildup in the liver and arteries. Other ingredients in oral chelation formulas include iodine, inositol, PABA, folic acid, choline, thymus, flax powder, bromelain, gentian, selenium, chromium, and coenzyme Q10. There are several products available today, some formulated by respected researchers and experts on chelation therapy.

Arteriosclerosis: Cause and Mechanism

When the body's arteries harden making them less elastic and less able to handle the body's blood flow, the condition is known as arteriosclerosis. The decrease in elasticity, combined with the clogging that is brought about by built-up plaque, is what leads to heart attack, stroke and other serious conditions. But what causes arterial plaque to build up in the first place, and why do the arteries harden? Some may offer the simple answer that all "clogging" in the arteries in caused by dietary cholesterol, but research indicates that there are various and more complex causes. This section gives a detailed explanation of the causes of atherosclerotic buildup and other coronary heart diseases, clarifying why chelation therapy is such an attractive option for treating heart disease.

Buildup of Calcium/Cholesterol Plaque

One of the most prominent causes of arterial buildup (referred to as atherosclerosis) involves the body's various detoxification processes and organs. When we eat, drink, and breathe, the body uses what is necessary and then, through the lungs, liver, kidneys, skin and colon, eliminates those things that are unnecessary or harmful to the body. Often, however, these processes become overwhelmed by the huge number of impurities to which we expose ourselves.

It is this overwhelming number of toxins and impurities that leads to the buildup of arterial plaque. How? Dr. Hitendra Shah, MD, offers an explanation: "Degenerative cardiovascular disease is, simply put, caused by calcium deposits which occur on heart valves, the heart muscle, and artery walls. Calcium is always present in the bloodstream, for it's necessary to sustain life. But under certain circumstances, calcium deposits form in the heart or arteries and interfere with blood flow" (Walker 1997, 122). Dr. Shah explains that any time the body is injured, it sends calcium to the problem area because calcium helps the affected tissues to bind and ultimately heal. Says Dr. Shah: "If injuries are infrequent, this repair system functions well. However, when blood vessel walls, heart muscle or heart valves undergo chronic, repetitive injury, a dangerous situation exists. Then, each time the injury gets repeated, calcium floods into the damaged area, and after a series of such frequent injuries, the calcium builds up sufficiently to bring on arterial blockages" (Walker 1997, 122).

The "injury" Dr. Shah refers to warrants more detailed discussion—it is not necessarily a bloody wound or bruise caused by physical trauma, but rather consists of tiny openings in the artery wall lining caused by constant irritation. The causes of this constant irritation include the following:

• Toxic materials contained in food, drugs, polluted air and water, second-hand cigarette smoke and many other sources that will usually end up in the bloodstream, especially if the body's eliminatory systems are overloaded. Constant exposure of the blood vessel walls to toxins will surely cause them to become sensitive and irritated. Dietary cholesterol has received much of the credit for being the chief irritant. In fact, when oxygen combines with cholesterol, it forms cholesterol oxide, a toxic form of cholesterol that is damaging to the blood vessel linings.

• A state of acidosis (too much acid) in the body can corrode the cell wall membranes in the heart, the arteries and the veins. Ultimately, cardiovascular structures and interconnective tissues are weakened. Overacidity also acts as another irritant to the vessel walls, instigating the process of calcium/cholesterol buildup. In fact, though cholesterol has long been thought to be the principal cause in plaque buildup, research shows that lowering cholesterol levels does not negate cholesterol plaque formation. Acidosis can also lead to the formation of free radicals. What causes acidosis? The major cause is an improper diet. The foods we eat largely determine acid levels in our bodies. Fatty and high-protein foods are the major culprits of acidosis, and these same fats (and cholesterol) usually compound the problem by being directly involved in arterial plaque formation (*Växa Journal*, 5–6).

• Free radicals are also being targeted as a main culprit in blood vessel irritation. Though free radical formation can be somewhat complex, a simple explanation is that a molecule can sometimes gain an unpaired electron and fall out of its electrical balance. In an attempt to stabilize itself, it seeks out "normal" or stable molecules for their electrons. If this occurs in the vessel wall lining, the lining becomes irritated. This is known as lipid peroxidation, and is believed by many experts to be a principal cause of atherosclerosis.

• More than fifty years ago, Johan Bjorksten, PhD, proposed the theory that premature aging occurs because of "cross-linking" between metallic minerals and proteins. His initial focus was that of aluminum and protein toxicity. Recently, his idea has been proven with findings that a combination of aluminum and beta amyloid protein is a principal source for the formation of Alzheimer's disease and other forms of dementia (Walker 1997, 120). Dr. Bjorksten's theory is well accepted by the scientific and medical communities.

But what exactly is cross-linking? Dr. Morton Walker, a chelation therapy pioneer explains:

> In the presence of toxic metal pollution, molecular chains of proteins, nucleic acids and polyvalent metals such as calcium, lead, aluminum and iron (which have more than one electrical charge in an atom) combine with other long protein molecules and form unnatural cross-links. Then the newly created, individual, giant cross-linked protein is no longer able to function normally. It cannot be split or be hydrolyzed, as is usually done by enzymes present in the blood system. Cross-linking produces free radicals that bring on pathology by preventing the usual splitting and hydrolization of proteins for use by the body. Thus, these free radicals from metallic minerals become highly destructive body toxins (Walker 1997, 120).

It is clear that atherosclerosis, or arterial buildup, comes about because of various factors. Toxins, free radicals, and dietary substances act as irritations to the blood vessel walls. In an attempt to heal these irritated areas, the body sends calcium to bind the area. The calcium then combines with other metals, fat, cholesterol and scar tissue and finally hardens into arterial plaque. If the body's circulatory system continually experiences this process, then chelation therapy can be an option in reversing cardiovascular disease.

Calcium Complexes and the Hardening of Arteries

Of the minerals most involved in the calcification of the blood vessel system, calcium (specifically ionic calcium) is the most damaging. Ionic calcium is a "floating" form of calcium that the body uses in daily functions like muscle contraction and relaxation, nerve impulse transmission, blood coagulation and others. However, as explained earlier, calcium is a mineral capable of forming complexes with other components, including proteins. Such complexes eventually lead to the formation of lesions, plaque, and the overall hardening of the blood vessels. The following list includes the major components (found mainly in arterial walls) that combine with calcium.

- **Elastin** is a type of protein that makes up a substantial percentage of the blood vessel wall. Just as its name suggests, elastin is the substance that allows the arterial wall to be elastic. During the processes that lead to atherosclerosis, elastin forms complexes with ionic calcium and as a result loses its elasticity.

- **Collagen** is another type of protein that works with elastin to make up the bulk of the arterial walls. It forms complexes with ionic calcium, leading to hardening of the blood vessel.

- **MPCs (mucopolysaccharides)** are carbohydrates that contain a number of agents (amino acids, uronic acids, and chondroitin sulfate, among others), all found within the arterial wall. They form complexes with ionic calcium to promote atherosclerosis.

- **Beta lipoproteins and pre-beta lipoproteins** (also known as low density lipoproteins and very low density lipoproteins) transport the combination of fatty acids and glycerol for storage in the liver, muscles and other areas of the body. (See Dr. Morton Walker's *Everything You Should Know about Chelation Therapy* for more information.)

While beta and pre-beta lipoproteins form ionic calcium complexes and instigate the onset of arteriosclerosis, there are also other lipoproteins (high density lipoproteins and very high density lipoproteins) that do not form complexes with calcium but actually interfere with the formation of the ionic calcium complexes.

Clearly, ionic calcium plays an important role in the formation of arterial plaque and the actual hardening of the arteries because of complexes it forms with arterial wall components. And because EDTA effectively ties up calcium and is eliminated via the urine, it is clear why EDTA chelation therapy would be successful in reducing the levels of atherosclerotic plaque and reversing the hardened condition of the artery walls.

Standard Treatments for Heart Disease

Coronary Artery Bypass Surgery

Coronary artery bypass graft surgery (CABG) can be a very risky procedure. It involves removing the portions of veins from the patient's legs and using them to circumvent, or bypass, the obstructed arteries that normally let the heart function. As stated earlier, such a procedure can involve some very serious health risks. First, there is a 25 percent chance that the patient will suffer some sort of heart attack while on the operation table, and under the worst circumstances, 10 to 17 percent of the patients die while undergoing the procedure. There can also be further deterioration of the bypassed arteries, with a substantial possibility of blockage of the new grafted blood vessels within two or three years. There is also a similar possibility of the reappearance, even within months, of chest pain, with little chance of relief by additional bypass or related surgery. Though the procedures involved in constructing a bypass system of foreign blood vessels have improved over the years, there is still substantial risk involved, not only the risk of suffering various complications, but also of death from the procedure.

Cost is another aspect of bypass surgery's controversial nature. At approximately fifty thousand dollars per procedure, it is no wonder it is such a popular recommendation of heart surgeons, doctors and specialists. Compared to the cost (time and financial) and success rate of chelation therapy, coronary bypass surgery certainly does not deserve to be the remedy of choice.

Angioplasty

The procedure percutaneous transluminal angioplasty (PTCA), otherwise known as angioplasty, involves inserting and inflating a balloon in the affected artery to press the built-up plaque against the walls of the artery, and possibly to "stretch" and encourage the artery to regain its lost elasticity. However, like coronary bypass surgeries, angioplasty procedures have posted only a mediocre record. Close to 40 percent of all angioplasties result in the operated artery simply closing up again. Half of repeat angioplasties must be done a third time. And a fifth of all patients undergoing angioplasties eventually receive coronary bypass surgery because their previous angioplasty (or angioplasties) fail. These numbers certainly are not encouraging; in fact, they are arguably less than acceptable.

In the attempt to make angioplasty a more effective treatment, doctors have developed a cross-wire mechanism that is inserted into the artery and left to prevent the arterial plaque from thickening and the artery from closing in on itself. However, the "improvement" hasn't been shown to be any more effective than the original angioplasty procedure (Walker 1997, 208). Comparing all aspects involved in receiving angioplasty surgery—money, recovery time, and most importantly, effectiveness in reversing the condition—it is very clear chelation therapy is a much more favorable route to take than angioplasty surgery.

History of EDTA and Chelation Therapy

The German chemist F. Munz is largely credited with synthesizing EDTA. It was patented in Germany in 1930 and first used in 1941 for lead poisoning. It was then patented in the U.S. in 1949, with several papers published on its therapeutic effects in the early 1950s.

EDTA chelation therapy has been employed in the U.S. to treat atherosclerosis since 1952, but was also used before that for lead poisoning and heavy metal toxicity. (Its use for lead poisoning is currently approved by the Food and Drug Administration). Soon after its initial use for lead and heavy metal poisoning, it was noted that EDTA chelation therapy resulted in reduction of angina pectoris (severe pressure and pain in and around the chest). It was this observation that initiated the investigation of EDTA to treat atherosclerotic vascular disease. One of the first studies reported on the results of EDTA treatment in twenty patients suffering from atherosclerosis, nineteen of whom were found to have substantial improvement. The authors state: "A placebo action seemed improbable for several reasons. There was slight if any clinical improvement until after the discomfort of twenty, and with a few, thirty infusions. A patient's progress was based on measured physical activity rather than appraisal of subjective symptoms. The improvement gained has been retained and expanded in all, and a few for as long as two years. The results have been uniformly good" (Clark, et al., 665).

A study conducted by Meltzer, et al., showed very favorable results concerning EDTA chelation treatments for atherosclerosis. Although the subjects showed minimal improvement after the initial session, nine of the ten patients reported significant improvement as measured by a decrease in the number and severity of angina attacks, reduced nitroglycerin use, increased work capacities and, in eight of the patients, improvement in their ECG results.

Thousands of scientific articles have been written concerning the various aspects of EDTA chelation therapies, and its safety has been proven by its use on hundreds of thousands of patients in well over three million separate intravenous treatments by over 1,000 doctors over the last fifty years (Walker 1982, 14). And as far as can be determined, not one fatality has been documented in contemporary chelation therapy when the established protocol has been followed. In fact, in approving the recent investigatory new drug application (IND) for EDTA, the FDA did not require that any additional safety studies be conducted to determine its safe use.

Why Is Chelation Not a Common Medical Practice?

Chelation therapy is largely unknown, ignored, and condemned by the conventional medical establishment. Organizations like the American Heart Association and the American Medical Association have openly denounced chelation therapy as ineffective and unproven. Why is this? There are several reasons, most of which are politically influenced.

The biggest reason chelation therapy is not endorsed by the conventional medical establishment—not a threat of corruption or a threat to patients' health, but a threat of depriving that establishment of millions of dollars each year. The various organizations and individuals who play a part in treating cardiovascular health are paid tens of billions of dollars annually. Each bypass surgery performed brings in between $25,000 to $50,000, and each angioplasty up to $15,000. The numerous drugs for lowering cholesterol, normalizing heart rhythm, relieving pain, and reducing high blood pressure, combined with the numerous consultations, tests, retests and exams, amount to hundreds of millions of dollars. Because it poses an economic threat, chelation therapy brings on itself the derision and condemnation of the conventional medical world.

Pharmaceutical companies are also behind this effort to keep EDTA chelation therapy from becoming mainstream. There are several reasons for this. First, because EDTA chelation is now out of patent, no pharmaceutical company wants to undertake the huge sums of money it would take to adequately document the therapeutic effects of chelation for purposes of being recognized by the FDA. Second, because there are hundreds of products developed for treating the symptoms of heart disease—hypertension, angina, arteriosclerosis, etc.—developing a product that could prevent or at least improve a condition of heart disease would only mean that the need for those other hundreds of products would disappear, thereby resulting in a huge financial loss for these companies (Walker 1997, 242-45).

Dr. Morton Walker, in his book *Everything You Should Know about Chelation Therapy*, gives a thorough discussion of how chelation therapy has been unfairly treated by various groups of power. The point is that despite the apparent safety and effectiveness of chelation therapy in preventing and reversing the terrible effects of heart disease, it is being slighted by these groups for reasons that have only to do with money and politics. This is an unfortunate situation for those who suffer from heart disease, considering that doctors, pharmaceutical groups and others that make up the medical establishment should only have their patients' health in mind, not how much money they could lose by suggesting a treatment option like chelation therapy.

Research on Chelation Therapy

As stated earlier, L. Terry Chappell has been a primary proponent of EDTA and chelation therapy for degenerative vascular diseases. A 1993 meta-analysis conducted by Chappell and Stahl of 41 studies of EDTA chelation treatments in patients with vascular disease showed that chelation therapy was a very successful treatment. Overall, 87 percent of the patients had measurable improvement with treatment, as measured by ECG, ankle/brachial index, walking distance, exercise activity, Doppler testing, and others (Chappell, Stahl, 139–60). Several of the studies included in their analysis are significant. A discussion of their results follows.

A 1988 study, conducted by Olszewer and Carter, involved 2,870 patients who were suffering from various chronic degenerative diseases, but primarily vascular disease. Nearly 90 percent of the patients showed good to excellent improvement, measured primarily by walking distance, ECG and Doppler changes. This same group also completed a small blinded, crossover study with peripheral vascular sufferers who, after treatment with EDTA chelation therapy, showed particularly significant improvement in both walking distance and ankle/brachial index measurements (Olszewer, Carter, 41-9).

Results of a 1994 study by van Rij and associates comparing EDTA and a thiamin/ascorbate treatment for peripheral vascular disease showed EDTA had a 60 percent improvement rate. Most importantly, the patients in this study who underwent the EDTA treatments improved significantly in five of the studied parameters including resting ankle/brachial indices in both better and worse legs, two different types of physical exercise and femoral pulsatility indices. Also significant to this study is the fact that 26 percent of the EDTA chelation group achieved greater than 100 percent improvement in walking distance, compared with only 12 percent of the other control group. Among the nonsmokers and those who had quit smoking (smokers reportedly can't benefit from EDTA chelation therapy as well as nonsmokers), 66 percent of the EDTA group bettered their walking distance by 86 percent, but only 45 percent of the control subjects bettered their distances with an average of 56 percent improvement (van Rij, et al. 1194–99).

In 1993, Hancke and Flytlie treated 65 patients with EDTA chelation therapy who were waiting for bypass surgery for an average of 6 months. Eighty-nine percent of these were able to cancel their surgery because of significant improvement in their symptoms. Hancke and Flytlie also chelated 27 subjects who had been recommended for amputation—24 of these were able to forego amputation and save their limbs. Considering this, Chappell estimates that if similar results were obtained in the U.S., over 350,000 of 400,000 bypasses could have been avoided and 102,000 limbs could have been saved with the employment of EDTA chelation treatment in such patients in 1992 alone. What would the savings have been from the avoidance of these procedures? Chappell estimates they could have been as much as $8 million (Chappell, 33).

Interview with Dr. Garry Gordon

A recent interview with Dr. Garry Gordon, MD, another proponent of EDTA chelation therapy, reveals some noteworthy information (excerpted from *Life Enhancement,* April 1997).

LE: Is there much documentation to support the use of EDTA in this way?

GG: There are 7,000 known articles in our possession about EDTA, and I am sure if I hired somebody and we really went to work, there would be another 3,000 that we haven't found... Today there is a group that has proposed that one of the biggest actions of EDTA could be its ability to tie up transition metals, such as copper and iron, which are known to catalyze free-radical activity. That's very exciting stuff, and it is really deep when you get into advances in free radical pathology. When I first got into the field 27 years ago and asserted that calcium was a part of the atherosclerotic process, it was believed by my opponents that calcium was a late and unimportant part of the process.... Calcium is now being accepted as an excellent marker [of the onset of atherosclerosis], and not a late marker at all.

LE: So, with a given patient, when would you do IV chelation and when would you do oral chelation?

GG: Basically, my position is this: the oral is an insurance policy to guarantee that you stay alive long enough to take the IV when and if you choose to do so. I tell everybody that the IV produces a "youthifying" effect to deal with several basic aspects of aging, including the calcification of your blood vessels.

People who are symptomatic really have no choice. Once they have any symptoms of heart disease, or fail a treadmill test, or have silent ischemia picked up on a holter monitor, I tell them they have to take IV and stay on oral, because taking the oral too means the beneficial effect of the IV will not be lost rapidly.

I have taken on patients who were inoperable, who had already had every known form of bypass surgery, until there weren't any more veins in their legs to strip out to put in their heart. They were sent home to die, and I could get those people back to full functioning. I've had doctor friends who wouldn't take the IV at first, but who now are on oral EDTA, and are able to pass a treadmill stress test that they couldn't pass for 5 years. I've had lots of good things happen with oral EDTA-based supplement programs.

Other Benefits of EDTA Chelation Therapy

Arthritis

EDTA chelation therapy has been shown to be effective in relieving symptoms of arthritis, especially rheumatoid arthritis. Dr. Morton Walker cites various reasons for this. It normalizes the production of mucopolysaccharide, a complex of proteins and sugars which, when after its metabolism becomes disrupted, causes disorders of the connective tissues and joints. Arthritis is, of course, one of these (Walker, 1997, 74–75).

Hypertension

Walker also points to EDTA chelation as being an effective way of controlling high blood pressure. He cites a variety of reasons, including an increase of cadmium from kidney cells, the decreased peripheral resistance after removal of calcium and plaque, and increased serum magnesium (Walker, 1997, 75).

Diabetes

Diabetes mellitus has long been recognized as a cause in atherosclerosis and impaired peripheral blood flow. Many doctors who give their patients EDTA chelation therapy have noted that insulin requirements among their diabetic patients decreases substantially after experiencing chelation therapy.

Most importantly for diabetics, the binding action of chelation therapy can prevent the onset of peripheral circulation impairment, most often seen in sight problems (macular degeneration of the retina) and disorders with the feet and hands (numbness, tingling, and even gangrene, which requires amputation).

What Are the Costs of Chelation Therapy?

The financial costs of intravenous chelation therapy vary, mainly by geographical location. The important thing to remember is that chelation therapy costs much less than any other therapy that will treat the cardiovascular diseases discussed here. Some of these price differences have to do with services apart from the actual EDTA chelation infusion. Some patients will require doses of vitamins and minerals as part of their treatment because they are deficient in one or several. Also, various tests are done—usually on hair, blood, urine and saliva—in order to determine these deficiencies. These tests also incur costs. Diet analysis may also be done, and this can be very helpful in determining what foods you aren't eating, but should be, as well as other alterations you should probably make in your diet to take complete advantage of the chelation therapy. Anyone thinking of doing chelation should realize that a complete course of chelation therapy usually involves about twenty separate sessions. This is usually the minimum number of sessions for mild cardiovascular disease. Preventive sessions may be fewer in number.

Oral chelation costs are somewhat different. Because oral chelation pills are usually recommended as more a preventative measure rather than a way to quickly reverse advanced cardiovascular disease, they require long-term continual consumption. But their cost is usually considerably lower than that of intravenous chelation therapy,. Again, this is cheaper than IV chelation sessions, but it must be remembered that oral products are meant to be taken continually as a preventative measure against the onset of cardiovascular disease.

How Safe Is Chelation Therapy?

Besides vascular diseases and heavy metal poisoning, EDTA has been shown to help battle scleroderma, rheumatoid arthritis, and osteoarthritis. It is safer than other chelating agents that have been used for rheumatoid disease, with studies also showing it chelates with unwanted calcium deposited in the areas of the joints, tendons and muscles.

One may wonder about the possibility of chelation therapy causing a lack of calcium in the bones and structural system of the body, ultimately causing osteoporosis or a similar disorder. The answer to this is a strong no. The reason for this is that the form of calcium used by the body—called calcium apatite—is different from that bound by EDTA, and virtually impossible to be chelated by something like EDTA. Several studies targeting the possibility that EDTA causes bone calcium loss have indicated that it does not (Walker 1997, 121).

Overall, EDTA chelation therapy has been shown to be very safe. Virtually no side effects have been discovered, and its long-term use has not resulted in any documented severe side effects. However, like most therapies, there are some individuals who should consult their health practitioner before undergoing EDTA chelation therapy. Pregnant women should avoid it, unless poisoning by heavy metal poses a more formidable risk. An individual with impaired kidney function should also consult their doctor. Also, long-term use of EDTA could deplete an individual of his/her trace element levels. This can be remedied by taking a multivitamin/mineral supplement before and after each chelation session (see Dr. Morton Walker's *Everything You Should Know about Chelation Therapy* for more detailed information). Occasionally, an individual may be mildly allergic to EDTA, but this is rare. The general consensus among those doctors practicing chelation therapy (both intravenous and oral) is that it is a safe and effective therapy for combating the effects of cardiovascular disease and other disorders.

References

"10 Reasons to Avoid an Acid pH." *Växa Journal 2* (1997): 5–6.

Chappell, L. Terry, MD. "Chelation Therapy, Smoking and Health Care Costs." *Journal of Advanced Medicine 7* (1994): 107.

Chappell, L. T., Michael Janson, MD. "EDTA Chelation Therapy in the Treatment of Vascular Disease." *Journal of Cardiovascular Nursing 10* (1996): 78–86.

Chappell, L. T., J. P. Stahl. "The Correlation between EDTA Chelation Therapy and Improvement in Cardiovascular Function: a Meta-Analysis." *Journal of Advanced Medicine 6* (1993): 139–60.

Clark, N. E., et al. "Treatment of Angina Pectoris with Disodium Ethylene Diamine Tetraacetic Acid." *American Journal of Medical Science December* (1956): 654–66.

Dean, Ward. "Dr. Ward Dean Comments on Chelation Therapy." *Life Enhancement April* (1997): 13.

Gordon, Garry, MD, DO. Interview with Life Enhancement Magazine. "Oral Chelation for Improved Heart Function." *Life Enhancement April* (1997): 8–15.

Hancke, C., K. Flytlie. "Benefits of EDTA Chelation Therapy on Arteriosclerosis." *Journal of Advanced Medicine 6* (1993): 161–72.

Meltzer, L. E., et al. "The Treatment of Coronary Artery Heart Disease with Disodium EDTA." In *Metal-Binding in Medicine*. M. Seven, ed. Philadelphia, Penn.: J. B. Lippincott, 1960.

Olszewer, E. J. Carter. "EDTA Chelation Therapy in Chronic Degenerative Disease." *Medical Hypotheses 27* (1988): 41–9.

Patterson, E. "Coronary Vascular Injury Following Transient Coronary Artery Occlusion: Prevention by Pre-Treatment with Deferoxamine, Dimethylthiourea and N-2 Mercapto Proprionyl Glycine." *Journal of Pharm. Exp. Ther. 266* (1993): 1528–35.

van Rij, A., et al. "Chelation Therapy for Intermittent Claudication: A Double-Blind, Randomized, Controlled Trial." *Circulation 90* (1994): 1194–99.

Walker, Morton, DPM. *The Chelation Answer*. New York: M. Evans and Co., 1982.

Walker, Morton, DPM. *Everything You Should Know about Chelation Therapy*. New Canaan, Ct.: Keats, 1997.

Healthy Reading!